U.S. SPECIAL OPS FORCES

INSIDE  DELTA FORCE

HOWARD PHILLIPS

PowerKiDS press

NEW YORK

Published in 2022 by The Rosen Publishing Group, Inc.
29 East 21st Street, New York, NY 10010

First Edition

Editor: Greg Roza
Designer: Rachel Rising

Portions of this work were originally authored by Jeanne Nagle and published as *Delta Force*. All new material in this edition was authored by Howard Phillips.

Photo Credits: Cover Oleg Zabielin/500px/500Px Plus/Getty Images; cover, pp. 1, 3, 4, 6, 7, 8, 9, 10, 12, 13, 14, 16, 17, 18, 20, 21, 22, 24, 25, 26, 27, 28, 30, 31, 32 CRVL/Shutterstock.com; cover, p.1 Zsschreiner/Shutterstock.com; cover, pp. 1, 3, 26, 27, 28, 30, 31, 32 Massimo Saivezzo/Shutterstock.com; p. 4 guvendemir/E+/Getty Images; p. 5 Spanic/E+/Getty Images; pp. 6, 7 Stocktrek Images/Getty Images; pp. 7, 9, 13, 17, 21, 25 Khvost/Shutterstock.com; p. 8 https://commons.wikimedia.org/wiki/File:UK_SAS_(badge).svg; p. 9 Will McIntyre/The LIFE Images Collection/Getty Images; p. 11 miodrag ignjatovic/E+/Getty Images; p. 13 U.S. Army/Handout/Getty Images News/Getty Images; p. 15 PRESSLAB/Shutterstock.com; p. 17 MICHEL GANGNE/Staff/AFP/Getty Images; p. 18 John Moore/Staff/Getty Images News/Getty Images; p. 19 Patrick A. Albright/U.S. Army; p. 20 Chmiel/iStock Unreleased/Getty Images; p. 21 Davis Turner/Stringer/ Getty Images News/Getty Images; p. 23 South_agency/E+/Getty Images; p. 25 SOPA Images/Contributor/LightRocket/Getty Images; p. 27 IROZ GAIZKA/Contributor/AFP/Getty Images; p. 29 Joe Raedle/Staff/Getty Images News/Getty Images.

Some of the images in this book illustrate individuals who are models. The depictions do not imply actual situations or events.

Library of Congress Cataloging-in-Publication Data

Names: Phillips, Howard, 1971- author.
Title: Inside Delta Force / Howard Phillips.
Description: New York : PowerKids Press, [2022] | Series: U.S. Special Ops forces | Includes bibliographic references ans index.
Identifiers: LCCN 2020036523 | ISBN 9781725328792 (library binding) | ISBN 9781725328778 (paperback) | ISBN 9781725328785 (6 pack)
Subjects: LCSH: United States. Army. Delta Force–Juvenile literature. | Special forces (Military science)–United States–Juvenile literature.
Classification: LCC UA34.S64 P54 2022 | DDC 356/.1670973-dc23
LC record available at https://lccn.loc.gov/2020036523

Manufactured in the United States of America

CPSIA Compliance Information: Batch #BSPK22. For further information contact Rosen Publishing, New York, New York at 1-800-237-9932.

CONTENTS

DELTA WHAT?

The United States military is among the largest and most powerful fighting forces in the world. Out of the more than 1 million soldiers on active duty, there are certain **elite** groups tasked with the most dangerous and difficult missions. The Navy SEALs are one such group, as are the Green Berets and the U.S. Special Forces Operational Detachment-Delta—more commonly known as Delta Force.

Each of the roughly 1,000 members of Delta Force is one of the best soldiers the country has to offer. The group's operations are often in the news. However, there's little public information about this group and its members.

ALL SPECIAL FORCES, INCLUDING DELTA FORCE, ARE **COVERT**. WORKING IN SECRECY IS PART OF THEIR SUCCESS.

The members of Delta Force are called operators rather than soldiers. Since this is a covert force, the number of active-duty members is classified. In addition to those who are responsible for actual fighting—called assault forces—there are medical, **intelligence**, training, and other support teams.

Delta Force operators are organized into several assault and support **squadrons**. Within each squadron, there are smaller groups called troops. Each troop includes teams that specialize in certain skills, including parachuting, scuba diving, and sharpshooting. Delta Force operators must be well rounded in all areas of warfare and ready to jump into danger at a moment's notice.

DELTA WOMEN

THE ORIGINAL DELTA FORCE OPERATORS WERE ALL MEN, AND SO ARE MOST OF TODAY'S MEMBERS. IN THE 1990S, HOWEVER, DELTA FORCE MADE IT A PRIORITY TO **RECRUIT** SOME FEMALE MEMBERS. THERE HAVE BEEN SOME WOMEN IN THE UNIT EVER SINCE. MANY WORK IN DELTA FORCE'S INTELLIGENCE DIVISION, WHICH IS NICKNAMED THE "FUNNY **PLATOON**."

ALL DELTA FORCE OPERATORS MUST GO THROUGH PARACHUTE TRAINING.

THE HISTORY OF DELTA FORCE

The 1970s were a dangerous time around the world. **Terrorist** groups became highly active, with members **hijacking** planes and killing or kidnapping political figures. The U.S. military saw the need to create a group of elite soldiers who specialized in fighting terrorism.

Delta Force was formed in 1977. Its founder, U.S. Army colonel Charles Beckwith, picked the unit's first recruits from a group of **volunteers**. To make the cut, soldiers had to pass a challenging physical fitness test, including dozens of push-ups and sit-ups, as well as long-distance runs. Recruits trained in crossing difficult **terrain** while carrying heavy packs.

UK SPECIAL AIR SERVICE REGIMENT

UK INSPIRATION

DURING HIS MILITARY CAREER, CHARLES BECKWITH SPENT A YEAR WORKING WITH THE BRITISH SPECIAL AIR SERVICE (SAS). FORMED IN 1941 DURING WORLD WAR II, THE SAS FIGHTS TERRORISTS, RESCUES **HOSTAGES**, AND ENGAGES ENEMY TROOPS. COLONEL BECKWITH CREATED AND ORGANIZED DELTA FORCE TO BE SIMILAR TO THE SAS.

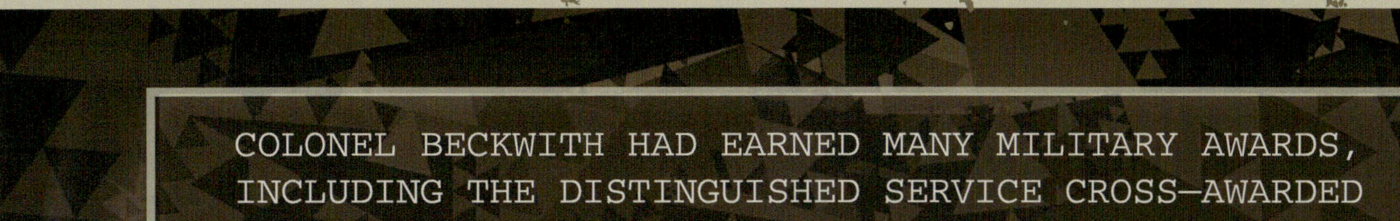

COLONEL BECKWITH HAD EARNED MANY MILITARY AWARDS, INCLUDING THE DISTINGUISHED SERVICE CROSS—AWARDED FOR HEROIC BRAVERY IN BATTLE—AND THE PURPLE HEART MEDAL FOR BEING INJURED IN BATTLE.

A handful of the early volunteers were chosen to join the new unit. Many of those who weren't selected either failed the physical tests or felt the pressure was too great and quit before the selection process ended.

After months of preparation, the original unit's final test is thought to have taken place at a remote North Carolina location. The operators worked as a team to enter both a building and an airplane to rescue pretend hostages. After showing their skills as covert operatives, Delta Force members were ready to get to work.

DELTA FORCE MEMBERS PARTICIPATE IN SHARED TRAINING PROGRAMS WITH GROUPS AROUND THE WORLD, SUCH AS BRITISH AND AUSTRALIAN SAS UNITS AND ISRAEL'S UNIT 269.

THE WAR ON TERROR

Today, Delta Force's main responsibility is counterterrorism. This means fighting against and trying to stop terrorist activity.

Counterterrorism efforts include two parts, and Delta Force handles both. The first is to counterattack after a terrorist act occurs. For instance, after the attack against the United States on September 11, 2001, U.S. Special Forces and other arms of the U.S. military were deployed to countries where the terrorists came from. The second part is to stop terrorism before it happens. This includes locating the hiding places of people believed to be terrorists and eliminating terrorist leaders.

BEYOND WAR

AS PART OF THE U.S. MILITARY, DELTA FORCE IS GENERALLY NOT PART OF MISSIONS INVOLVING REGULAR PEOPLE INSIDE THE UNITED STATES. HOWEVER, THE UNIT HELPED THE FBI STOP A 1987 RIOT IN A GEORGIA PRISON. IN 1993, THE UNIT ALSO SET UP **SURVEILLANCE** DURING A POLICE STANDOFF WITH A GROUP IN WACO, TEXAS.

SEPTEMBER 11, 2001, WAS ONE OF THE WORST DAYS IN AMERICAN HISTORY, AND IT HIGHLIGHTED THE NEED FOR UNITS LIKE DELTA FORCE.

HELPING HOSTAGES

In addition to counterterrorism, Delta Force operators also rescue people being held hostage overseas. In fact, the unit's first mission, in 1980, was to free Americans being held against their will in Iran. This mission failed, but since then, the unit has carried out rescue operations around the world.

Successfully rescuing hostages is very hard, even for the best of the best. Rescuers often have to use force to enter the place where hostages are being held. During a fight, it can be hard to tell who needs to be saved. Delta Force operators are trained to keep hostages safe while they deal with kidnappers.

HOSTAGE RESCUE MISSIONS ARE SOME OF THE MOST DANGEROUS AND DIFFICULT WORK DELTA FORCE DOES.

KEEPING PEOPLE SAFE

The thought of becoming a Delta Force operator may seem like signing up for an action film. However, some Delta Force jobs aren't very exciting, but they are very important. For example, the unit is sometimes responsible for protecting VIPs at home and abroad. Operators may protect top U.S. military commanders and leaders visiting from other countries.

Delta Force has also provided security for special events with large international audiences. One of the unit's first assignments was to protect players and fans at the 1979 Pan-American Games. Delta Force also provided security for the Olympic Games in Los Angeles (1984) and Atlanta (1996).

DELTA IN THE GULF

DURING THE PERSIAN GULF WAR OF 1991, MEMBERS OF DELTA FORCE SERVED AS BODYGUARDS FOR GENERAL NORMAN SCHWARZKOPF, THE COMMANDER OF 700,000 U.S. AND INTERNATIONAL SOLDIERS IN IRAQ. OTHER UNIT MEMBERS WHO WEREN'T INVOLVED WITH SECURITY DURING THE WAR WENT ON SECRET MISSIONS TO FIND ENEMY WEAPONS.

GENERAL SCHWARZKOPF (CENTER) IS SHOWN EXITING A HOTEL IN RIYADH, SAUDI ARABIA, IN 1991. DELTA FORCE OPERATORS SURROUND HIM, READY FOR ANYTHING.

WHAT IT TAKES

Delta Force is one of the Army's most respected units. Each year, many soldiers either volunteer or are chosen to become members. A lot of these soldiers come from elite units within the U.S. Army, such as the Rangers and the Green Berets. However, any member of the U.S. armed forces can volunteer for Delta Force duty.

In order to make the cut in this unit, soldiers must first make it through a difficult and lengthy selection process. During this time, their skills with guns and advanced weapons are tested. So are their physical fitness and mental toughness. The selection process lasts an entire month.

DELTA FORCE RECRUITS MUST BE EXPERIENCED SOLDIERS WITH TOP-NOTCH SHOOTING SKILLS, OUTSTANDING PHYSICAL FITNESS, AND A GOOD MENTAL PROFILE.

Soldiers who make it through the selection process then start their training to become Delta Force operators. This takes place at the Delta Force headquarters: the Fort Bragg military base in North Carolina. Delta Force shares Fort Bragg with other Special Forces units.

New members of Delta Force are trained in all the skills necessary for modern counterterrorism. For several months, recruits practice target shooting, handling explosives, parachuting out of airplanes, climbing up and down tall buildings, high-speed driving, scuba diving, and hand-to-hand combat. They also receive training in surveillance and other intelligence work.

NOT FOR THE FAINT OF HEART

THE "HOUSE OF HORRORS" IS A DELTA FORCE TRAINING BUILDING. IT REPORTEDLY CONTAINS ALL KINDS OF VEHICLES IN WHICH THE UNIT CAN STAGE FAKE HOSTAGE RESCUES. RECRUITS OFTEN RUN TRAINING COURSES WITH REAL BULLETS—AND OFTEN WITH REAL PEOPLE ACTING AS HOSTAGES!

FORT BRAGG IS HOME TO SOME OF THE MOST ELITE MILITARY UNITS IN THE UNITED STATES. HERE, PRESIDENT OBAMA SPEAKS TO SOLDIERS AT FORT BRAGG IN 2011.

ELITE EQUIPMENT

As one of the country's top military forces, Delta Force provides its operators with a huge variety of weapons and gear. Some of these items are standard issue, meaning they're used widely across the military. An example of a standard-issue weapon is the M4, a rifle most soldiers carry. Bulletproof vests are worn by all troops fighting on the front lines.

Other weapons are used mainly by special ops forces. For instance, lightweight machine guns and rifle-mounted **grenade** launchers are often issued. Delta Force operators also have access to the most advanced military technology for gathering intelligence and cyberwarfare.

MANY OF DELTA FORCE'S WEAPONS ARE DESIGNED TO HAVE MULTIPLE PURPOSES, FROM FIREFIGHTS TO BREAKING INTO BUILDINGS.

ON THE MOVE

Delta operators commonly drive a type of four-wheel drive vehicle called a Humvee. First used by the U.S. military in the 1980s, fully equipped Humvees come with machine guns and armor that can protect occupants from explosives and gunfire.

Another military group–the 160th Special Operations Aviation Regiment (SOAR)–often flies operators to mission locations. Operators parachute in from low-flying planes and helicopters. If they can't find a place to land, operators may rescue people using a rope lowered from a helicopter.

Delta Force missions sometimes require water travel. Advanced submarines and military speedboats help Delta Force get where they need to go quickly and quietly.

FAST ATTACK

IN THE TOUGH TERRAIN IN WHICH DELTA FORCE MEMBERS ARE OFTEN DEPLOYED, THEY RELY ON DESERT PATROL VEHICLES (DPVs) TO GET AROUND. THESE LIGHTLY ARMORED DUNE BUGGIES ARE EASY TO DRIVE ON SAND AND OVER ROCKS. DELTA FORCE AND THE NAVY SEALS USED THESE VEHICLES, FORMERLY CALLED FAST ATTACK VEHICLES (FAVS), DURING WARS IN THE PERSIAN GULF.

THE HUMVEE IS WIDELY USED IN THE U.S. MILITARY BECAUSE IT OFFERS MOBILE FIREPOWER AND PROTECTION.

DELTA WORK DONE

The following are some of the most famous operations in the history of Delta Force.

1980: OPERATION EAGLE CLAW

✪ An attempt to rescue hostages in Iran. Unfortunately, it was a failure. A helicopter and plane crashed and burned before the hostages could be rescued.

1989: OPERATION JUST CAUSE

✪ As part of a complex military mission in Panama, Delta Force operators helped locate a military **dictator** named Manuel Noriega, who was ruling through fear and violence.

1993: OPERATION GOTHIC SERPENT

✪ Delta Force members in Mogadishu, Somalia, were tasked with capturing former Somali clan leader Muhammad Farah Aydid. Aydid was targeted after he killed United Nations troops helping people in the country.

NO MATTER WHERE THEY ARE SENT, DELTA FORCE OPERATORS ARRIVE ARMED WITH HIGH-TECH WEAPONS AND PROTECTIVE GEAR.

2016: OPERATION BLACK SWAN

✪ While on the hunt for a killer connected to a drug gang, Delta Force operators found—and captured—Joaquín "El Chapo" Guzmán, one of the world's most powerful drug lords.

2019: OPERATION KAYLA MUELLER

✪ On this mission, named after an American held hostage by a terrorist group, Delta Force operators fought and killed Abu Bakr al-Baghdadi, a terrorist leader in Iraq.

The public will likely never know how many missions Delta Force has undertaken. As with many covert military units, the U.S. government only reveals a small amount of information about its activities and objectives. Nonetheless, Delta Force operators are happy to work in silence, protecting their country and the world.

DELTA FORCE OPERATORS RISK THEIR LIVES AND WORK TIRELESSLY TO PROTECT AMERICANS AROUND THE WORLD.

GLOSSARY

covert: Made or done secretly.

dictator: A leader who has total power and often limits citizens' rights.

elite: Superior in talent and ability.

grenade: A small bomb thrown by hand or by a launcher.

hijack: To take control of a vehicle by force.

hostage: A person who is captured by someone who demands certain things before freeing the captured person.

intelligence: The gathering of secret information about enemies.

platoon: A small group of soldiers who work together.

recruit: To invite people to join a military force. Also, a new member of a military force.

squadron: A division of military organization, smaller than a unit but larger than a platoon.

surveillance: The act of watching someone or something closely.

terrain: A type of land in an area.

terrorist: A person who uses violence to scare people as a way of achieving a political goal.

volunteer: A person who offers to do something.

FOR MORE INFORMATION

BOOKS

Haskell, L. S. *Missions of the Delta Force*. Mankato, MN: The Child's World, 2016.

Perritano, John. *A Terrorist Goes Down! Delta Forces in Syria Take Out an ISIS Leader*. Broomall, PA: Mason Crest, 2019.

WEBSITES

How Delta Force Works

science.howstuffworks.com/delta-force.htm

This detailed website has more information about the history and makeup of Delta Force.

U.S. Army Special Forces

www.goarmy.com/special-forces.html

Learn more about U.S. Army Special Forces and how soldiers become a part of them.

Publisher's note to educators and parents: Our editors have carefully reviewed these websites to ensure that they are suitable for students. Many websites change frequently, however, and we cannot guarantee that a site's future contents will continue to meet our high standards of quality and educational value. Be advised that students should be closely supervised whenever they access the internet.

INDEX